Bibliographic information published by the German National Library:

The German National Library lists this publication in the National Bibliography; detailed bibliographic data are available on the Internet at http://dnb.dnb.de .

Imprint:

Copyright © 2016 GRIN Verlag, Open Publishing GmbH
Print and binding: Books on Demand GmbH, Norderstedt Germany
ISBN: 9783668439146

This book at GRIN:

http://www.grin.com/en/e-book/356941/renewable-energy-resource-potentials-and-constraints-in-nigeria

Luqman Adedokun

Renewable Energy Resource Potentials and Constraints in Nigeria

GRIN Publishing

GRIN - Your knowledge has value

Since its foundation in 1998, GRIN has specialized in publishing academic texts by students, college teachers and other academics as e-book and printed book. The website www.grin.com is an ideal platform for presenting term papers, final papers, scientific essays, dissertations and specialist books.

Visit us on the internet:

http://www.grin.com/

http://www.facebook.com/grincom

http://www.twitter.com/grin_com

RENEWABLE ENERGY RESOURCE POTENTIALS AND CONSTRAINTS IN NIGERIA: AN APPRAISAL.

BY:

ADEDOKUN LUQMAN ADEBAYO[1]

Abstract

The effect of the use of fossil fuel in our daily activities has adverse effects on our society. As a result, the need to make a shift to other sources of Energy is paramount. Climate Change in the world we live in is no longer news, thus the need to seek alternative sources of energy to ensure that the earth remains habitable for mankind. The Nigerian environment affords the exploitation of various renewable source of energy, ranging from solar energy, wind, tidal, just to mention a few. Nigeria has the opportunity to harness its energy resources in a clean and environmentally friendly manner. This paper discusses The Renewable Energy Resources in Nigeria, its sources, features, potentials and constraints.

---

[1] PhD Student in the field of Environmental Law and Sustainable Development at the Atlantic International University, Honolulu, Hawai, United States.

Content

1.0 Introduction

1.1 Definition of Renewable Energy.

There is much debate about how to define and distinguish renewable energy from non-renewable energy forms, and the terms and definitions chosen can have huge impacts on policy and regulatory efforts aiming to promote clean energy resources. **Texas Renewable Energy Industry Alliance (TREIA's)** definition of renewable energy hereinafter reproduced is hereby adopted.

> "Renewable energy: Any energy resource that is naturally regenerated over a short time scale and derived directly from the sun (such as thermal, photochemical, and photoelectric), indirectly from the sun (such as wind, hydropower, and photosynthetic energy stored in biomass), or from other natural movements and mechanisms of the environment (such as geothermal and tidal energy). Renewable energy does not include energy resources derived from fossil fuels, waste products from fossil sources, or waste products from inorganic sources."[2]

Renewable energy is that form of energy obtained from sources that are essentially inexhaustible, unlimited and rapidly replenished or naturally renewable such as wind, water, sun, wave, refuse, biofuels etc.[3]

Renewable energy is that which is replaced naturally or controlled carefully and can therefore be used without the risk of finishing it all.

Renewable energy came to the limelight of the global discourse and attention due to increasing and devastating effects that the predominate use of fossil energy resources caused on Climate change.

---

[2] http://www.treia.org/renewable-energy-defined/ Retrieved on 04/10/2016
[3] Etiosa Uyigue (Executive Director Community Research and Development Centre) Welcome address on Promoting Renewable Energy and Energy Efficiency in Nigeria. Held at the University of Calabar Hotel and Conference Centre 21st November 2007.

2.0. Renewable Energy Sources in Nigeria

2.1. Solar Energy

2.2 Solar energy, an energy obtained from the sun, is the world's most abundant and cheapest source of energy available from nature. Solar energy is available in two forms, namely Solar Thermal and Solar PV.[4] Nigeria receives abundant solar energy that can be usefully harnessed with an annual average daily solar radiation of about 5.25 kW h/m2 /day. This varies between 3.5 kW h/ m2 /day at the coastal areas and 7 kW h/m2 /day at the northern boundary. The average amount of sunshine hours all over the country is estimated to be about 6.5 h. This gives an average annual solar energy intensity of 1,934.5 kW h/m2 /year; thus, over the course of a year, an average of 6,372,613 PJ/year (approximately 1,770 TW h/year) of solar energy falls on the entire land area of Nigeria.[5] This is about 120,000times the total annual average electrical energy generated by the Power Holding Company of Nigeria (PHCN). With a 10% conservative conversion efficiency, the available solar energy resource is about 23 times the Energy Commission of Nigeria's (ECN) projection of the total final energy demand for Nigeria in the year 2030.[6]

2.3 It is also estimated that the technical potential of solar energy in Nigeria with 5% device conversion efficiency put at 15.0 × 1014 kJ of useful energy annually. This equates to about 258.62 million barrels of oil equivalent annually, which corresponds to the current national annual fossil fuel production in the country. This will also amount to about 4.2 × 105 GW/h of electricity production annually, which is about 26 times the recent annual electricity production of 16,000 GW/h in the country.[7] However, to enhance the developmental trend in the country, there is every need to support the existing unreliable energy sector with a sustainable source of power supply through solar energy.[8]

[4] C. N. Ezugwu. Renewable Energy Resources in Nigeria: Sources, Problems and Prospects. Journal of Clean Energy Technologies, Vol. 3, No. 1, January 2015. http://www.jocet.org/papers/171-E2003.pdf retrieved on 04/02/17

[5] Chineke TC, Igwiro EC (2008) Urban and rural electrification: enhancing the energy sector in Nigeria using photovoltaic technology. African Journal Science and Tech 9(1):102–108

[6] Energy Commission of Nigeria (ECN) (2005) Renewable Energy Master Plan

[7] Onyebuchi EI (1989) Alternative energy strategies for the developing world's domestic use: A case study of Nigerian household's final use patterns and preferences. The Energy Journal 10(3):121–138

[8] Oyedepo Energy, Sustainability and Society 2012, 2:15 http://www.energsustainsoc.com/conten. Retrieved on 04/10/2015

2.4 Wind Energy

Many indigenous researchers have also explored the availability of wind energy sources in Nigeria with a view of implementing them if there is likelihood for their usage. The wind speed data of 30 stations in Nigeria was analysed to determine the annual mean wind speeds and power flux densities, which vary from 1.5 to 4.1 m/s to 5.7 to 22.5 W/m2, respectively.[9]

This report points to the fact that the nation is blessed with a vast opportunity for harnessing wind for electricity production, particularly at the core northern states, the mountainous parts of the central and eastern states, and also the offshore areas, where wind is abundantly available throughout the year.

2.5 Hydro Energy

Hydropower supplies about 40% of the total electric power in Nigeria.[10] Its primary source is large rivers such as Niger and Benue and their several tributaries and natural falls possessing high hydropower potentials for the country. Nigeria has exploited its hydro energy sector, e.g. the kanji dam. It is reported that the total hydroelectric power potential of the country is estimated to be about 8,824 MW with an annual electricity generation potential in excess of 36,000 GW h. This consists of 8,000 MW of large hydropower technology, while the remaining 824 MW is still small-scale hydropower technology.[11]

3.0 Features of Renewable Energy

3.1 Nigeria is blessed with a large amount of renewable natural resources, which, when fully developed and utilized, will lead to poverty reduction and sustainable development. The following features are common to the Renewable Energy sources.

- It is free, since it comes from the sun, the wind, waves etc.
- The implementation of renewable energy technologies will help to address the environmental concerns that emerged due to greenhouse gas emissions such as carbon

---

[9] Adekoya LO, Adewale AA (1992) Wind energy potential of Nigeria. Renewable Energy 2(1):35–39
[10] Awogbemi, O. and Komolafe, C.A. (2011).Potential for Sustainable Renewable Energy Development in Nigeria. The Pacific Journal of Science and Technology, 12, 161-169.
[11] Akinbami JFK (2001) Renewable Energy Resources and Technologies in Nigeria: Present Situation, Future Prospects and Policy Framework'. Mitigation and Adaptation Strategies for Global Change 6:155–181. Kluwer Academic Publishers, Netherlands 14. Akinbami JFK, Ilori M.O, Oyebisi T.O.

dioxide (CO2), oxides of nitrogen (NOx), oxides of sulfur (SOx), and particulate matters as a result of power generation from oil, natural gas, and coal. In the long term it will mitigate climate change.

- The renewable energy sources cannot be depleted. If used carefully in appropriate applications, renewable energy resources can provide a reliable and sustainable supply of energy almost indefinitely. In contrast, fossil fuel resources are diminished by extraction and consumption.
- Its facilities require less maintenance than traditional generators. Their fuel being derived from natural and available resources reduces the cost of operation.
- Economic benefits for rural areas.

4.0 Renewable Energy Prospects in Nigeria.

4.1 The Prospects of renewable energy in Nigeria are enormous; it will not only boost the economic development of Nigeria, but will also ensure Energy Security in Nigeria. Research works in this area should be encouraged to enable provision of sufficient and reliable energies for most communities in the country.[12]

4.2 The cost-competitiveness of renewable energy gives the country the opportunity to dramatically increase its ambition and demonstrate the industry's financial viability in the region, while also securing a stable and very low-risk supply of energy, thereby extending the lifetime of its fossil fuel reserves.[13] Renewable Energy will, to a large extent free up the Petroleum and Gas Industry and therefore enables the country to generate significant new export revenues.

Priority in Renewable Energy in Nigeria, especially in the area of research will show Nigeria's commitment to sustainable development on a global scale. Most of the research should be targeted at improving efficiency and increasing overall energy yields.[14]

---

[12] C. N. Ezugwu. Renewable Energy Resources in Nigeria: Sources, Problems and Prospects. Journal of Clean Energy Technologies, Vol. 3, No. 1, January 2015. http://www.jocet.org/papers/171-E2003.pdf retrieved on 04/10/15

[13] Masdar Institute/IRENA (2015), Renewable Energy Prospects: United Arab Emirates, REmap 2030 analysis. IRENA, Abu Dhabi. www.irena.org/remap retrieved on 04/06/2016.

[14] S. C. E. Jupe, A. Michiorri, and P. C. Taylor, "Increasing the energy yield of generation from new and renewable energy sources," Renewable Energy, vol. 14, no. 2, pp. 37-62, 2007.

5.0 Constraints of Renewable Energy in Nigeria

5.1 Despite the good prospects for the development of renewable energy resources in Nigeria, there are still some challenges militating or that are potentially capable of militating against the development of renewable energy resources in the country.

5.2 First, the high costs of executing renewable energy projects relative to other sources of energy significantly affects the economics of renewable energy projects such that it is difficult for them to break even on a commercial basis. This is a major disincentive to developing renewable energy projects on a commercial basis in Nigeria.

5.3 Secondly, in spite of the formulation of the National Energy Policy, 2003 which addresses several energy subsector including different renewable energy sources, there is currently no law that addresses issues of renewable energy. The reality is that specific legal and regulatory intervention is required in order to alleviate some of the impediments to developing renewable energy projects. One area in which this intervention is needed is in the taxation of renewable energy projects. Currently, renewable energy projects are subject to the same tax regime as any other energy development projects except the oil and gas upstream energy industry. The implication is that given the financing constraints associated with renewable energy projects, significant tax incentives need to be given in order to enhance the economics of such projects.

5.4 The other constraints common to Renewable Energy Supply in Nigeria are as follows:

- Renewable Energy is perhaps most hamstrung by lack of governance structures that consider net costs and benefits.
- Supply of renewable energy is oftentimes unreliable. Renewable energy often relies on the weather for its source of power. For instance, hydro generators need rains to fill reservoirs of dams to enable sufficient head of water that can generate high power.[15] Also, wind turbines require wind to turn the blades, and solar collectors need clear sides and sunshine to collect heat and make electricity. When these resources are not available, so is the capacity to make energy from them. This can be unpredictable and inconsistent.

---

[15] Same as footnote 11.

- The increase in Energy demand might not be well catered for, due to storage difficulties of renewable energy.

- The current cost of renewable energy technology is also far in excess of traditional fossil fuel generation. The reason is that it is a new technology having extremely large capital cost.

- Renewable Energy may be discounted because they cannot provide the same consistency of 24-hour power supply or on-and-off options.

6.0 Nigeria's Current Situation on Renewable Energy

6.1 In realization of energy deficiencies in Nigeria and the global trend towards clean environment, the Federal Government of Nigeria decided to take issues of alternative energy supply seriously in order to reduce the reliance on oil and gas as the major source of energy in the country. Given recent global trend in the energy industry the obvious alternative is renewable energy. In this regard, the Federal Government of Nigeria issued the National Energy Policy, 2003 to set out the policy objectives of the Federal Government of Nigeria regarding the entire energy sector and the various strategies to help in achieving those objectives. While the National Energy Policy document contained policy objectives for the entire energy sectors, it addresses each subsector including the renewable energy subsector.
The formulation of the policy objectives for renewable is a major step towards boosting the quota of renewable energy to the total energy mix in Nigeria. This is because the formulation of these policies has set out a clear agenda of direction for government actions. For example, the policy objectives will create a clear direction for the formulation of appropriate laws and regulatory regime to address the challenges associated with the development of renewable energy resources in Nigeria.

6.2 In addition, there is huge gap between energy demand and energy supply in Nigeria creating an uncommon level of energy deficiencies. This supply-demand imbalance creates a strong economic factor for renewable energy development projects in Nigeria notwithstanding the high cost associated with many renewable energy projects.

6.3 Conclusively, despite several efforts being made by Nigeria to improve compliance with internationally acceptable environmental standards, the country is still far cry from achieving these standards. For example, Nigeria continues to be a major gas flaring nation in spite of all steps that have been taken over the years to reduce gas flaring. It is also common knowledge that the enforcement drive of the government towards ensuring the compliance by oil and gas companies to comply with environmental laws and regulations is not encouraging. All these make the development of renewable energy resources an irresistible option to pursue.

7.0 Recommendations

7.1 There is a strong prospect of renewable energy in Nigeria making significant contribution to boosting economic development of Nigeria. As renewable energy projects can be executed in small scales, it can provide a good source of energy for many micro, small and medium scale enterprises which have been acknowledged as the back bone of poverty alleviation and economic development.

7.2 The development of renewable energy resources will also ensure Energy Security in Nigeria because the abundant renewable energy resources can make significant addition to energy supply in Nigeria towards bridging the demand-supply imbalance in the energy industry. For example, it is a very good option for rural electrification and off grid areas in Nigeria. This is recognized under **section 88 (9) (c) of the Power Sector Reforms Act, 2005** which requires the Minister to include information on renewable energy in its quarterly report the Minister is required to submit to the President on rural electrification.

7.3 Research works in this area should be encouraged to enable provision of sufficient and reliable energies for most communities in the country.[16]

7.4 Priority in Renewable Energy in Nigeria, especially in the area of research will show Nigeria's commitment to sustainable development on a global scale. Most of the research should be targeted at improving efficiency and increasing overall energy yields.[17]

---

[16] C. N. Ezugwu. Renewable Energy Resources in Nigeria: Sources, Problems and Prospects. Journal of Clean Energy Technologies, Vol. 3, No. 1, January 2015. http://www.jocet.org/papers/171-E2003.pdf retrieved on 04/10/15

[17] S. C. E. Jupe, A. Michiorri, and P. C. Taylor, "Increasing the energy yield of generation from new and renewable energy sources," Renewable Energy, vol. 14, no. 2, pp. 37-62, 2007.

8.0   Conclusion

8.1 A clear deployment programme would be a catalyst for renewable energy and likely further reduce costs. After determining economic and environmental attractiveness, energy authorities would be able to reap the benefits by announcing and following an investment regime.

8.2 A policy framework for distributed generation could mobilize private investment and social support for renewable energy.

8.3 Conclusively, the federal government should partner with the private sector to make these alternative energy sources available and affordable to most communities in our country.

References:

1. Oxford Advanced Learner"s Dictionary New 8th Edition.

2. Akinbami JFK (2001) Renewable Energy Resources and Technologies in Nigeria: Present Situation, Future Prospects and Policy Framework'. Mitigation and Adaptation Strategies for Global Change 6:155–181. Kluwer Academic Publishers, Netherlands 14. Akinbami JFK, Ilori MO, Oyebisi T.O, Akin.

3. C. N. Ezugwu. Renewable Energy Resources in Nigeria: Sources, Problems and Prospects. Journal of Clean Energy Technologies, Vol. 3, No. 1, January 2015.

4. S. C. E. Jupe, A. Michiorri, and P. C. Taylor, "Increasing the energy yield of generation from new and renewable energy sources," Renewable Energy, vol. 14, no. 2, pp. 37-62, 2007.

5. Awogbemi, O. and Komolafe, C.A. (2011).Potential for sustainable renewable energy development in Nigeria. The Pacific Journal of Science and Technology, 12, 161-169.

6. Chineke TC, Igwiro EC (2008) Urban and rural electrification: enhancing the energy sector in Nigeria using photovoltaic technology. African Journal Science and Tech 9(1):102–108.

7. Onyebuchi EI (1989) Alternative energy strategies for the developing world's domestic use: A case study of Nigerian household's final use patterns and preferences. The Energy Journal 10(3):121–138.

8. Adekoya L.O, Adewale A.A (1992) Wind energy potential of Nigeria. Renewable Energy 2(1):35–39.

9. Energy Commission of Nigeria (ECN) (2005) Renewable Energy Master Plan

10. Etiosa Uyigue (Executive Director Community Research and Development Centre) Welcome address on Promoting Renewable Energy and Energy Efficiency in Nigeria. Held at the University of Calabar Hotel and Conference Centre 21st November 2007.

11. Oyedepo Energy, Sustainability and Society 2012, 2:15

12. Masdar Institute/IRENA (2015), Renewable Energy Prospects: United Arab Emirates, REmap 2030 analysis. IRENA, Abu Dhabi.

.

Websites

http://wwhsdc.org/wp-content/uploads/2015/03/IJISBT-49-72.Onifade.pdf.

http://www.nnpcgroup.com/NNPCBusiness/MidstreamVentures/RenewableEnergy.aspx.

http://www.jocet.org/papers/171-E2003.pdf

www.irena.org/remap

http://www.jocet.org/papers/171-E2003.pdf

http://www.energsustainsoc.com/conten.

http://www.treia.org/renewable-energy-defined/

YOUR KNOWLEDGE HAS VALUE

- We will publish your bachelor's and
  master's thesis, essays and papers

- Your own eBook and book -
  sold worldwide in all relevant shops

- Earn money with each sale

Upload your text at www.GRIN.com
and publish for free